I0467984

Information Technology/ Information Management

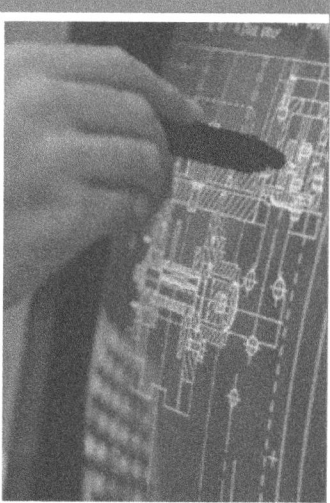

Strategic Plan
Fiscal Years 2008–2013

U.S.NRC

United States Nuclear Regulatory Commission

Protecting People and the Environment

A MESSAGE FROM THE CHIEF INFORMATION OFFICER

Darren B. Ash

I am pleased to issue this update to the U.S. Nuclear Regulatory Commission's (NRC's) Information Technology/Information Management (IT/IM) Strategic Plan. The objective of the NRC's IT/IM program is to—

> Manage agency information and employ information technology to improve the productivity, effectiveness, and efficiency of agency programs, and enhance the availability and usefulness of information to all users inside and outside the agency.

Through its achievement we will help the NRC fulfill its mission:

> License and regulate the Nation's civilian use of byproduct, source, and special nuclear materials to ensure adequate protection of public health and safety, promote the common defense and security, and protect the environment.

The goals, strategies, and measures in this plan provide the foundation for directing and assessing the performance of the NRC's IT/IM program over the next 5 years. Its scope covers all of the NRC's IT/IM resources agencywide, including our local and wide area networks, computers, and telecommunication devices, our information and records management functions, and all of our applications ranging from mission-critical systems, such as those supporting licensing and emergency response, to systems required for support functions like payroll, personnel, and accounting.

The NRC updated this plan to maintain consistency with the NRC Strategic Plan, a new version of which was issued in February 2008, as well as to place additional emphasis on IT security and make improvements to the IT/IM performance measures. The NRC will report the performance for key measures to Congress, beginning with the results for FY 2009.

The most important goal of the NRC's IT/IM program remains the same—to improve access to NRC information for both staff and stakeholders. This goal continues to drive our other IT/IM goals, which have been modified to place additional emphasis on IT security. The other five goals now cover IT applications, IT security, IT infrastructure, human capital, and customer service. The next 5 years will present many challenges for the NRC's IT/IM program—guiding the modernization of the NRC's IT infrastructure and legacy applications, supporting the licensing of new reactors, and helping the agency to operate with a larger, more dispersed, and more mobile workforce while also protecting NRC's information and information technology in an increasingly complex threat environment.

As Chief Information Officer, I am accountable for the agency's IT/IM program, but its scope ultimately extends to every NRC organization and affects all of our stakeholders. Because of that, you—our stakeholders, both inside and outside the agency—continue to be a key element of this plan. As part of the implementation process, we will be asking for your input as we move forward with many of our strategies and measures, and I can assure you that your input is and will continue to be a key driver of the agency's IT/IM program.

Table of Contents

Executive Summary

The U.S. Nuclear Regulatory Commission's Information Technology/Information Management Strategic Plan for Fiscal Years 2008–2013 (NRC IT/IM Strategic Plan FY 2008–FY 2013) describes how IT/IM activities at the NRC help accomplish the agency's mission. That mission includes licensing and regulation of commercial nuclear power plants; research, test, and training reactors; nuclear fuel cycle facilities; medical, academic, and industrial uses of radioactive materials; and the transport, storage, and disposal of radioactive materials and wastes. The NRC's regulations are designed to protect the public and occupational workers from radiation hazards in those industries using radioactive materials.

The NRC's agency Strategic Plan for FY 2008 –FY 2013 contains the following strategy for IT/IM in the "Operational Excellence" section under "Organizational Excellence": "Manage information and employ information technology to improve the productivity, effectiveness, and efficiency of agency programs and enhance the availability and usefulness of information to all users inside and outside the agency." The IT/IM Strategic Plan adopts this strategy as its overall strategic objective and lays out the IT/IM goals, strategies, and performance measures needed to achieve it. It should be noted that the IT/IM Strategic Plan also supports strategies in the "Openness" and "Effectiveness" sections of the NRC Strategic Plan as well as strategies associated with the NRC's Safety and Security goals.

The IT/IM plan responds to Federal requirements laid out in the Paperwork Reduction and Clinger-Cohen Acts, which, among other things, direct agencies to establish goals and measures of the contribution of IT/IM activities to agency productivity, efficiency, effectiveness, and service to the public.

The IT/IM strategic planning process began with a situation assessment covering the needs of NRC stakeholders, NRC's major program drivers, environmental factors, and the strengths and weaknesses of the NRC IT/IM program.

Carried out with the participation of program officials from across the agency, the planning process resulted in the IT/IM program objective, vision, and strategic goals shown in the box at right, as well as high-level strategies and measures for each of the strategic goals.

The NRC IT/IM strategic planning process is fully integrated with the NRC's agencywide planning, budgeting, and performance management process and establishes clear accountability for the agency's IT/IM strategies and measures.

In summary, the goals, strategies, and measures in the NRC IT/IM Strategic Plan provide the foundation for directing and assessing the performance and results of the NRC's IT/IM program over the next 3 to 5 years.

The NRC's IT/IM Program

Objective: Manage information and employ information technology to improve the productivity, effectiveness, and efficiency of agency programs and enhance the availability and usefulness of information to all users inside and outside the agency

Vision: Getting the right information to the right people at the right time, efficiently and effectively

Strategic Goals

- *Goal 1—Information: Make it easy for the staff to produce and access information to perform their work and for stakeholders to participate and interact effectively with the agency*

- *Goal 2—IT Applications: Achieve and sustain effective, easy-to-use, and integrated IT applications that support the management of information throughout its lifecycle*

- *Goal 3—IT Security: Protect the NRC's information and information systems to ensure their integrity, confidentiality, and availability*

- *Goal 4—IT Infrastructure: Provide an IT/IM infrastructure that is secure, robust, reliable, and responsive to changing business needs*

- *Goal 5—IT/IM Human Capital: Recruit, develop, and retain a highly capable IT/IM workforce with the competencies needed to support NRC's goals and objectives*

- *Goal 6—IT/IM Customer Service: Achieve and sustain a high level of satisfaction with agencywide information services*

1. Introduction

The U.S. Nuclear Regulatory Commission's Information Technology/Information Management Strategic Plan for Fiscal Years 2008–2013 (NRC IT/IM Strategic Plan FY 2008–FY 2013) describes how IT/IM activities at the NRC contribute to the agency's mission. Section 1 covers the NRC's mission and responsibilities, Federal requirements for IT/IM planning, benefits of IT/IM strategic planning, and the relationship of this plan to other planning documents. Section 2 provides the environment and context for the plan. Sections 3 and 4 discuss the planning outcomes—the IT/IM program objective, vision, and strategic goals, and the strategies and measures associated with each of the goals. Section 5 describes the relationship of the IT/IM strategic planning process to other agency planning processes and to the Federal Electronic Government Program.

1.1 About the NRC

The NRC was established by the Energy Reorganization Act of 1974 and began operations in 1975. Its mission is to license and regulate the Nation's civilian use of byproduct, source, and special nuclear materials to ensure adequate protection of public health and safety, promote the common defense and security, and protect the environment.

The NRC is responsible for licensing and regulating commercial nuclear power plants; research, test, and training reactors; nuclear fuel cycle facilities; medical, academic, and industrial uses of radioactive materials; and the transport, storage, and disposal of radioactive materials and wastes. The NRC's regulations are designed to protect the public and occupational workers from radiation hazards in those industries using radioactive materials. For more information about the NRC's activities, see http://www.nrc.gov.

1.2 Federal IT/IM Strategic Planning Requirements

The Paperwork Reduction Act and the Clinger-Cohen Act of 1996 set forth the Federal requirements associated with IT/IM strategic planning. This document serves as the NRC's strategic information resources management plan in accordance with Section 3506(b)(2) of the Paperwork Reduction Act. The IT/IM strategic goals in Section 3.2 and performance measures in Section 4, together with performance measures for individual major IT investments, address the requirements of Section 3506(b)(3) of the Paperwork Reduction Act and Sections 5123(1) and (3) of the Clinger-Cohen Act. The NRC's annual budget submittal to Congress, beginning with the FY 2009 budget, will include annual progress in achieving the goals, as required by Section 5123(2) of the Clinger-Cohen Act.

1.3 Benefits of IT/IM Strategic Planning

Why did Congress create a requirement for IT/IM strategic planning? As early as 1994, the General Accounting Office (now the Government Accountability Office (GAO)) published a best practices report entitled, "Improving Mission Performance Through Strategic Information Management and Technology." Based on a series of case studies, the report laid out 11 fundamental strategic management practices that led to significant performance improvements, both short and long term, in leading private and public organizations. The report and subsequent GAO testimony before Congress was so compelling that these practices ultimately became the basis for the Clinger-Cohen Act. Several of the practices were directed towards strategic planning, including "Anchor strategic planning in customer needs and mission goals" and "Integrate the planning, budgeting, and evaluation processes." The NRC has made these two practices the cornerstones of the agency's new IT/IM strategic planning process.

Some other benefits of strategic planning include the following:

- Forces a look at the future and therefore the opportunity to influence it.
- Provides a sense of direction and continuity.
- Provides a framework for organizational decisionmaking.
- Results in the development of specific goals, strategies, and measures that provide standards of accountability for people, programs, and allocated resources.
- Aligns the total organization— people, processes, and resources— with a clear, compelling, and desired future state.

The NRC sees strategic planning as an essential element for the success of its programs. Because of the ever-increasing importance of information access and IT in agency business processes and the significance of IT/IM investments, it is essential to have a roadmap for IT/IM strategies and investments that moves the NRC, efficiently and effectively, in the direction set by the agency Strategic Plan. That roadmap is the NRC IT/IM Strategic Plan.

1.4 Relationship to Other Documents

The NRC IT/IM Strategic Plan is aligned with the NRC Strategic Plan and the NRC Performance Budget, which capture the NRC's planning, budgeting, and performance management processes:

- NRC Strategic Plan

The NRC Strategic Plan documents agency-level goals and strategies to meet the agency's mission. The NRC IT/IM Strategic Plan is consistent with the agency Strategic Plan and focuses more specifically on the IT and IM goals and strategies needed for the agency to accomplish its mission. Section 3.3 provides more information on the relationship between the two plans.

- NRC Performance Budget

The NRC Performance Budget provides the proposed outcomes and measures associated with the funding needed to meet the agency mission. Each FY, the NRC submits its budget to the Office of Management and Budget (OMB) and, later, to the Congress. The targets for key measures laid out in Section 4 of this IT/IM plan are included in the NRC Performance Budget and the NRC's IT/IM investments are aligned with the strategies and measures in this plan. The resource implications of the strategies and measures are determined as part of the budget process when targets are set. The NRC will report the performance for key IT/IM measures to Congress, beginning with the results for FY 2009.

2. Environment and Context

2.1 The NRC Strategic Plan

The NRC Strategic Plan forms the foundation for IT/IM strategic planning by laying out the NRC's overall strategic direction for the planning period. The plan lays out the following mission, values, strategic goals, associated outcomes, and organizational excellence objectives:

Mission

License and regulate the Nation's civilian use of byproduct, source, and special nuclear materials to ensure adequate protection of public health and safety, promote the common defense and security, and protect the environment.

Values

The safe use of radioactive materials and nuclear fuels for beneficial civilian purposes is enabled by the agency's adherence to the principles of good regulation—independence, openness, efficiency, clarity and reliability. In addition, regulatory actions are effective, realistic, and timely.

Strategic Goals

* Safety: Ensure adequate protection of public health and safety and the environment.
* Security: Ensure adequate protection in the secure use and management of radioactive materials.

Strategic Outcomes:

Safety:

* Prevent the occurrence of any nuclear reactor accidents.
* Prevent the occurrence of any inadvertent criticality events.
* Prevent the occurrence of any acute radiation exposures resulting in fatalities.
* Prevent the occurrence of any releases of radioactive materials that result in significant radiation exposures.
* Prevent the occurrence of any releases of radioactive materials that cause significant adverse environmental impacts.

Security:

* Prevent any instances where licensed radioactive materials are used domestically in a manner hostile to the United States.

Organizational Excellence Objectives to Achieve the Strategic Goals:

Openness:

The NRC appropriately informs and involves stakeholders in the regulatory process.

Effectiveness:

NRC actions are high quality, efficient, timely, and realistic, to enable the safe and beneficial use of radioactive materials.

Operational Excellence:

NRC operations use effective business methods and solutions to achieve excellence in accomplishing the agency's mission.

The full text of the NRC's Strategic Plan may be found at http://www.nrc.gov/about-nrc/plans-performance.html. The goals and strategies in this IT/IM plan contribute to both the safety and security goals and to the organizational excellence objectives in the agency's Strategic Plan. See Section 3.3 for a detailed description of the linkages between the NRC Strategic Plan and the goals and strategies in this plan.

2.2 The NRC IT/IM Program Management Approach

The agency's Chief Information Officer (CIO), who reports directly to the NRC's Executive Director for Operations (EDO), manages the NRC's agencywide IT/IM program. Reporting to the CIO are the Office of Information Services (OIS) and the Computer Security Office (CSO). Figure 1 shows the main components of the OIS and CSO. OIS manages and operates the agency's IT infrastructure, provides information and records services, and coordinates programs to assist with the development and maintenance of the NRC's business applications. OIS also manages the agency's IT/IM strategic planning, capital planning, and enterprise architecture activities. CSO oversees the agency's IT security program, including policy, training, certification and accreditation of IT systems, and cyber security situational awareness. For additional information on the OIS and CSO functions, see http://www.nrc.gov/about-nrc/organization.html.

The NRC headquarters and regional offices also have certain responsibilities for portions of the IT/IM program, the most important of which is the sponsorship of major applications that support their business functions. Application sponsors develop the business case and manage the application systems development lifecycle process with varying levels of assistance from OIS, depending on office needs. The NRC regional offices manage their local IT infrastructure. The NRC intends to continue with this basic approach to the allocation of IT/IM responsibilities.

Figure 1

Executives and Organizations with Primary Responsibilities
for the NRC's Agencywide IT/IM Program

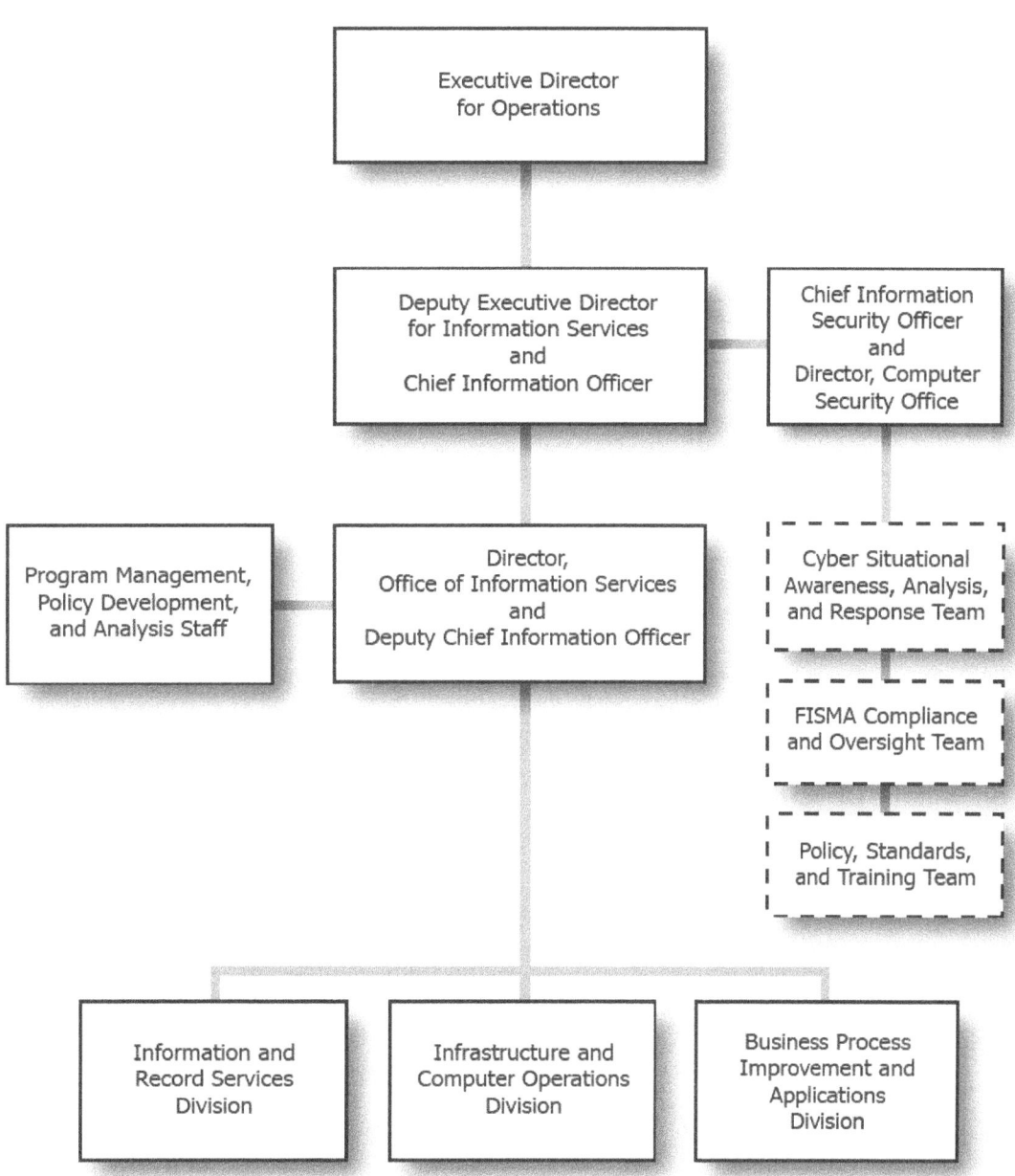

2.3 Situation Assessment

The IT/IM strategic planning process began with a situation assessment, reviewing the needs of NRC stakeholders and assessing internal program drivers; external oversight and Electronic Government drivers; the political, economic, and technological environment; strengths of the NRC IT/IM program; and areas for improvement based on recent self-assessments, Inspector General (IG) audits, and GAO reports.

The NRC's Stakeholders

The NRC's external stakeholders include the agency's licensees, the nuclear industry, advocacy organizations, the Congress, OMB, State and local governments, Indian Tribes, other Federal agencies (such as the Department of Energy (DOE), the Department of Homeland Security, and the Environmental Protection Agency), international nuclear regulators, academia, members of the public living near regulated facilities, and the general public.

Internal stakeholders include the Commission, senior executives, managers, and staff.

Each of these stakeholders has interests with respect to the NRC's activities. The stakeholder interests particularly relevant to IT/IM planning include the following:

- access by external stakeholders to agency information needed to understand the NRC's mission, goals, and performance and to participate effectively in the regulatory process
- access by internal stakeholders to agency information needed to perform their work, grow professionally, and take advantage of Federal employee services and benefits
- reliable IT infrastructure and high-quality IT/IM customer support services
- efficient IT governance processes to allow IT/IM projects to move forward expeditiously while ensuring that the IT/IM budget is invested wisely and that IT/IM investments produce the intended return
- protection of privacy data
- ability to communicate and share information (securely, in some cases) internally and with external stakeholders
- ability to conduct business effectively and efficiently, both internally and with external stakeholders
- need for more efficient business processes to handle an increased workload
- protection of stakeholder rights and agency accountability by keeping information as long as needed for legal, financial, and operational purposes, as well as identifying information needed for historical preservation

NRC Programmatic Drivers

Agency programs and activity areas expected to be the main drivers for new and existing IT applications during the planning period include the following:

- new reactor applications and licensing
- high-level waste repository licensing review, as affected by the uncertainty about the timing of the submittal of the DOE license application
- homeland security, including secure communication
- nuclear nonproliferation
- changes in incident response requirements and increased coordination with Federal response partners
- increased volume of electronic communication
- defense in depth for IT infrastructure and information security
- continued emphasis on the NRC's objective to ensure openness in its regulatory processes
- continued maintenance and modernization of major applications supporting NRC's regulatory processes

- enhancing the National Materials Program, which includes activities such as medical, industrial, and academic uses of radioactive materials; increased controls and tracking of radioactive sources; and implementation of the Energy Policy Act of 2005, mandating an NRC regulatory framework for certain naturally occurring and accelerator-produced radioactive material

- emergent work involving new fuel cycle technologies and industry initiatives to increase production

External Oversight and Electronic Government Drivers

External oversight and Electronic Government drivers include the following:

- opportunities to achieve efficiencies and increase openness through Electronic Government initiatives

- migration to fee-for-service shared administrative support systems under Electronic Government initiatives (such as requirements to use specific service providers for human resource and financial systems)

- increasing demands for information and heightened expectations of stakeholders with respect to Electronic Government driving the automation of business processes from end to end, including stakeholder business functions and public dissemination of information (e.g., the NRC's new processes for conducting hearings electronically)

- continued focus of oversight agencies on how agencies are meeting security and privacy requirements, including those in the Federal Information Security Management and Privacy Acts

- continued need for rapid response to Governmentwide initiatives with significant IT implications such as Homeland Security Presidential Directive 12 (HSPD-12) and Federal implementation of Internet Protocol Version 6 (IPV6)

- continued oversight focus from OMB based on the requirements of the Clinger-Cohen Act related to the effective management of IT investments from an enterprise perspective (project management, earned value management, capital planning and investment control (CPIC) process, portfolio management, etc.)

- the continued oversight focus from OMB based on the requirements of the Clinger-Cohen Act related to the use of enterprise architecture for better agencywide planning of IT investments, including common data, shared services, and business process orientation

- increased interest from the Congress and the Administration in how agencies are meeting their human capital challenges

- continued oversight focus from the National Archives and Records Administration (NARA) and OMB to comply with Federal records laws and regulations under 44 U.S.C. 2901, 3101, and 3102, as well as OMB Circular A-130, "Management of Federal Information Resources," related to establishing and maintaining continuous and systematic control of information throughout its lifecycle

Political Environment

The assessment considered the following aspects of the political environment:

- uncertainty about the timing of the high-level waste repository application

- elevated risk of terrorism and heightened security levels

- heightened concern about security and emergency preparedness as a result of terrorism, the threat of a flu pandemic, and recent natural disasters

- increased emphasis on Federal continuity of operations

Economic Environment and Workforce Changes

The assessment considered the following economic environment and workforce changes:

- a more computer-literate staff with less NRC corporate knowledge

- the Federal fiscal environment, providing a strong impetus for process efficiencies

- higher energy prices, space limitations, and staff retention needs driving an increase in telecommuting

- growth in the nuclear industry, increasing turnover and competition for qualified staff

Technological Environment

The assessment considered the following aspects of the technological environment:

- continuing advances in technology capabilities with processing power doubling every 18 months, storage capacity doubling every 12 months, and bandwidth doubling every 9 months, making it important to have an IT architecture, systems development, and governance approach that enables the NRC to respond to the rapid evolution of technology

- advances in mobile computing, storage, and wireless technologies, providing opportunities for innovation in regulatory and administrative processes

- advances in knowledge management and collaboration technologies providing opportunities to improve information access and information dissemination

- IT security moving from user input (user identification and password) to a two-factor security architecture (in which a user has something tangible to give access, such as a key, as well as information that is entered, such as user identification and password)

- increasing dependence on IT, heightening the need to prevent and combat cyberterrorism through strong computer security programs

- challenges stemming from the need to preserve information during an era of rapid technological change and storage media obsolescence

Strengths

A situation assessment usually includes a review of strengths and weaknesses, so that strengths can be built upon and weaknesses addressed. To prepare for this assessment, the NRC reviewed recent performance self-assessments, IG audits, GAO audits of the agency, and internal performance measures and assessments.

The NRC's IT/IM program was found to have strengths in several main areas. In the area of IM, a particular strength is the Document and Records Management and Information Collections Program. NARA has frequently recognized the NRC's records management program as a model for other Federal agencies. The NRC's Agencywide Documents Access and Management System (ADAMS) has the first NARA-approved records retention schedules for an enterprisewide electronic recordkeeping system. On May 13, 2003, NARA awarded the NRC the Archivist Achievement Award for Records Management in recognition of the deployment of ADAMS as a recordkeeping system. When NARA issued guidance to agencies for establishing a vital records program, it cited the NRC's program as an example for other agencies to follow.

Because of the NRC's exemplary information collections program and the quality of its submissions to OMB, the NRC has been granted special authorization to independently review and approve information collections with an insignificant burden impact on the public. The NRC Technical Library and the Public Document Room have also achieved recognition from their user groups for the depth of their collections and available resources, the high caliber of their professional staff, and the thoroughness of their services.

In the IT area, strengths include overall IT infrastructure reliability and support as well as special capabilities that have been put in place to support NRC hearings. The NRC's agencywide IT infrastructure and services support contract provides hardware/software installation, maintenance, and support; network infrastructure, maintenance, support, and administration; and central management of all desktop and network resources and services at headquarters, the regional offices, and resident inspector sites. In FY 2007, the NRC maintained an IT infrastructure availability service level of 99.9 percent and achieved a customer satisfaction score of 4.6 on a scale of 1 to 5.

Another IT strength is the special IT infrastructure that the agency has put in place for electronic hearings. The NRC's Electronic Information Exchange system allows those who do business with the agency to submit documents electronically with digital certificates. Using this system, along with the agency's document management system, electronic hearing dockets, and digital hearing management system, the NRC will be able to conduct its future hearings, such as those for new reactors and the high-level waste repository, electronically.

Areas for Improvement

Based on the performance review, the NRC's IT/IM program could benefit from improvements in the following areas:

- more effective guidance, support, and leadership for compliance with the Federal Information Security Management Act
- more effective communication and partnerships with customers and stakeholders to provide awareness of current and future initiatives, ensure alignment of priorities, effectively manage and coordinate service and work activities, and improve service delivery
- stronger leadership and clearer direction for IT/IM long-range planning, vision, and strategies
- more progress on the development and implementation of an agency enterprise architecture to more effectively plan and identify IT business solutions that support agency business needs
- more effective guidance, support, and leadership for IT investment management from an enterprise perspective
- improvements in training for IT project managers and staff with significant IT security responsibilities
- better approaches for email records management
- continued need to update processes and policies for the appropriate preservation of records and information, including partnering with industry and government to address challenges and develop solutions

3. IT/IM Program Objective, Vision, and Strategic Goals

3.1 IT/IM Objective and Vision

The box at the right shows the objective and vision of the NRC's agencywide IT/IM program. The NRC's IT/IM activities include all such activities across the agency, at headquarters as well as at the NRC's regional offices, resident inspector locations, and Technical Training Center. This includes all IT infrastructure, such as local and wide-area networks, desktop computers, and telecommunications; all application development and Web sites; and all IM activities across the agency, such as document management, records management, and Freedom of Information Act processing. Information utility is intended to include all aspects of usability, including information quality, timeliness, accuracy, availability, and ease of use.

3.2 IT/IM Strategic Goals

In response to the NRC's Strategic Plan and the situation assessment discussed in Section 2, the NRC has adopted six strategic goals for its IT/IM program:

(1) Information—Make it easy for the staff to produce and access information to perform their work and for stakeholders to participate and interact effectively with the agency.

(2) IT Applications—Achieve and sustain effective, easy-to-use, and integrated agency IT applications that support the management of information throughout its lifecycle.

(3) IT Security—Protect the NRC's information and information systems to ensure their integrity, confidentiality, and availability.

(4) IT Infrastructure—Provide an IT/IM infrastructure that is secure, robust, reliable, and responsive to changing business needs.

(5) IT/IM Human Capital—Recruit, develop, and retain a highly capable IT/IM workforce with the competencies needed to support NRC's goals and objectives.

(6) IT/IM Customer Service—Achieve and sustain a high level of satisfaction with agencywide information services.

These goals will be used to guide the NRC's IT/IM activities and investment priorities.

3.3 Relationship to Agency Strategic Plan

Section 2.1 of this document summarizes the NRC Strategic Plan, including its two strategic goals and three organizational excellence objectives. The IT/IM programs support all five of these elements. Table 1 maps some specific elements of the IT/IM Strategic Plan to the agency's strategic goals and organizational excellence objectives. The appendix provides a more exhaustive cross-reference between specific outcomes, strategies, and means for each goal in the NRC Strategic Plan and the six IT/IM strategic goals.

The NRC's IT/IM Program

Objective: Manage information and employ information technology to improve the productivity, effectiveness, and efficiency of agency programs, and enhance the availability and usefulness of information to all users inside and outside the agency

Vision:

Getting the right information to the right people at the right time, efficiently and effectively

Table 1

Examples of How the IT/IM Strategic Plan Supports the NRC Strategic Plan

Goals and Organizational Excellence Objectives with Associated Strategies from the NRC Strategic Plan	Supporting Goals and Strategies from the IT/IM Strategic Plan
Safety—Continue to oversee the safe operation of existing plants while preparing for and managing the review of applications for new power reactors.	IT Applications—Apply IT/IM to meet high-priority business needs (e.g., new reactors, the high-level waste repository proceeding, homeland security).
Safety—Effectively respond to events at NRC licensed facilities and other events of national interest, including maintaining and enhancing the NRC's critical incident response and communication capabilities.	IT Infrastructure—Strengthen IT infrastructure capabilities to accommodate agency business needs during emergencies.
Security—Use relevant intelligence information and security assessments to maintain realistic and effective security requirements and mitigation measures.	IT Security—Strengthen the security controls that protect NRC's IT systems and information using an efficient and effective certification and accreditation process.
Security—Share security information with appropriate stakeholders and international partners.	IT Security—Strengthen cyber security situational awareness and incident response.
Security—Control the handling and storage of sensitive security information, and the communication of information to licensees and Federal, State, and local partners.	Information—Improve electronic access to classified and safeguards information as appropriate to conduct agency business.
	IT Infrastructure—Provide IT infrastructure capabilities to accommodate classified and safeguards information.
Openness—Enhance the awareness of the NRC's independent role in protecting public health and safety, the environment, and the common defense and security.	Information—Improve information management processes such as information dissemination and knowledge management.
Openness—Provide accurate and timely information to the public about NRC's mission, regulatory activities, and performance and about the uses of, and risks associated with, radioactive materials.	Information—Improve internal and external electronic information access and delivery systems including ADAMS and the NRC's internal and public Web sites.
	Information—Improve NRC information quality, timeliness, and completeness.
Openness—Provide for fair, timely, and meaningful stakeholder involvement in NRC decisionmaking without disclosing classified, safeguards, proprietary, and sensitive unclassified information.	Information—Improve awareness of existing NRC information resources.
Effectiveness—Use state-of-the-art technologies and risk insights to improve the effectiveness and realism of NRC actions with a goal of continuous improvement.	IT Applications—Before implementing new IT applications, evaluate and improve associated regulatory and support processes, considering the needs of all process participants and using the most effective redesign approaches and technologies.
Effectiveness—Continue to improve the NRC's regulatory and communication programs.	IT Applications—Identify needs and provide necessary resources to deliver IT/IM services to all offices to help them take advantage of IT/IM capabilities to improve the effectiveness and efficiency of their operations.
Effectiveness—Achieve efficiencies in the licensing process that enable the safe and secure use of nuclear material.	

Goals and Organizational Excellence Objectives with Associated Strategies from the NRC Strategic Plan	Supporting Goals and Strategies from the IT/IM Strategic Plan
Operational Excellence—Improve support services to make them more efficient and make it easier to accomplish agency goals.	IT/IM Customer Service—Improve the effectiveness and efficiency of help services for the staff and the public.
Operational Excellence—Manage information and employ information technology to improve the productivity, effectiveness, and efficiency of agency programs and enhance the availability and usefulness of information to all users inside and outside the agency.	IT Applications—Seek common solutions, reduce duplication, and promote sharing of data, systems, and service components across the agency.
	IT Applications—Increase staff awareness, proficiency, and innovation in applying IT/IM tools to strengthen individual and organizational performance.
Operational Excellence—Use innovative strategies to recruit, develop, and retain a high-quality, diverse workforce.	IT Infrastructure—Increase IT infrastructure capacity, availability, and reliability to cost effectively meet business needs.
Operational Excellence—Provide accurate, timely, and useful financial information to agency managers for effective decisionmaking.	IT/IM Human Capital—Fully utilize available Federal competitive compensation and workforce flexibilities as well as specialized recruitment and development programs to attract and retain top level IT/IM talent.
	Information—Systematically assess staff and stakeholder information needs and develop plans to address them.

4. Strategies and Performance Measures by Goal

This section lays out a set of strategies and measures for achieving each of the six IT/IM strategic goals. The measures listed in the table for each goal relate to the associated goal rather than to a particular strategy. Each year during the agency budget process, the NRC will allocate resources for the specific projects (means) for implementing the strategies and set specific performance targets for each measure.

4.1 Goal 1: Information

Information: Make it easy for the staff to produce and access information to perform their work and for stakeholders to participate and interact effectively with the agency.

Information: Make it easy for the staff to produce and access information to perform their work and for stakeholders to participate and interact effectively with the agency.	
Strategies	**Performance Measures**
1. Systematically assess staff and stakeholder information needs and develop plans to address them.	1. Information Dissemination Timeliness—Meets agency timeliness targets for key information dissemination channels, including public meeting notices, Freedom of Information Act responses, and documents made publicly available through ADAMS.
2. Improve information and records management processes, such as information dissemination and knowledge management.	
3. Improve internal and external electronic information access and delivery systems including ADAMS and the NRC's internal and public Web sites.	2. External Stakeholder Satisfaction—Meets agency targets for external stakeholder satisfaction with key NRC information dissemination channels, including the NRC public Web site.
4. Improve NRC information quality, timeliness, and completeness.	3. Internal Stakeholder Satisfaction—Staff members' satisfaction with access to the information needed to do their job based on results from the NRC employee survey.
5. Improve awareness of existing NRC information resources.	
6. Improve electronic access to classified and safeguards information as appropriate to conduct agency business.	

The following section provides example means for executing each of the strategies that support the Information goal, followed by a brief discussion of the measures for this goal.

Strategy 1—Systematically assess staff and stakeholder information needs and develop plans to address them

The NRC will systematically identify, prioritize, and develop plans to address the information needs of major internal and external stakeholder groups. Example means for assessing user needs include surveying individual stakeholders, soliciting input from internal and external organizations, and obtaining input from NRC information system user groups. The results of this strategy and the two satisfaction measures listed in the table above will help to determine the focus areas for the other strategies associated with the Information goal.

Strategy 2—Improve information and records management processes, such as information dissemination and knowledge management

Based on the results of Strategy 1, the NRC will assess and improve IM processes such as the following:

- information dissemination
- methods for producing, collaborating on, and reusing information
- mechanisms for proper records identification and capture
- mechanisms for knowledge transfer and management

The NRC has reviewed its information dissemination program in accordance with OMB Memorandum M-06-02, "Improving Public Access to and Dissemination of Government Information and Using the Federal Enterprise Architecture Data Reference Model," dated December 16, 2005, and found that most NRC information is readily available through its public Web site (particularly the resources available in the Electronic Reading Room, which includes comprehensive access to all of the NRC's public documents), the services of the Public Document Room, and NUREG/BR-0010, Revision 4, "Citizen's Guide to the U.S. Nuclear Regulatory Commission," issued August 2004, which provides consistent, authoritative instructions on where and how to obtain NRC information for members of the public who may or may not have access to the Internet. The review did identify some areas for improvement, including the need to update the NRC's information dissemination policy and make it easier for the public to search the full scope of the NRC's public Web content. The NRC will use the areas for improvement identified by the review, together with findings from Strategy 1, the performance measurement process for Goal 1, and work associated with the NRC's data reference model, to continuously improve its information dissemination program.

Strategy 3—Improve internal and external electronic information access and delivery systems

Example means for executing this strategy include the following:

- making progress toward a coordinated agencywide search strategy encompassing all NRC electronic information sources
- applying best practice usability principles to make it easier for employees and the public to access NRC information from agency Web sites and application systems
- enhancing and upgrading the NRC's document and records management capabilities to make documents easier to find and access, and to accommodate new business requirements
- implementing internal knowledge management capabilities that facilitate information exchange within communities of interest

Strategy 4—Improve NRC information quality, timeliness, and completeness

The NRC will assess existing quality controls and, where appropriate, make improvements to ensure that information is complete, accurate, up to date, preserved (as appropriate), and made public in a timely manner.

Strategy 5—Improve awareness of existing NRC information resources

The NRC will facilitate internal and external stakeholders' understanding of information available and the mechanisms and resources that may be used to access or obtain the information.

Strategy 6—Improve electronic access to classified and safeguards information as appropriate to conduct agency business

Example means include the following:

- improving staff access to NRC safeguards and classified information by establishing electronic document and records management capabilities for this information, including secure electronic access by authorized users

- enabling more effective sharing of safeguards and classified information between the NRC and other agencies by participating in Federal initiatives such as the Homeland Secure Data Network

See also Strategy 8 under Goal 4, IT Infrastructure (Section 4.4).

Measures for Goal 1

The NRC has elected to use two measures of stakeholder satisfaction and one measure of information dissemination timeliness to assess performance in meeting the Information goal. The NRC will design surveys to provide actionable results that can help the agency to focus on the areas with the greatest impact.

4.2 Goal 2: IT Applications

IT Applications: Achieve and sustain effective, easy-to-use, integrated IT applications that support the management of information throughout its lifecycle.

Strategies	Performance Measures
1. Before implementing new IT applications, evaluate and improve associated business processes, considering the needs of all process participants and using the most effective redesign approaches and technologies.	1. Availability—Percent of identified applications that meet their specified availability target during the operating hours specified in service-level agreements.
2. Apply IT/IM to meet high-priority business needs (e.g., new reactors, the high-level waste repository proceeding, homeland security).	2. Staff Satisfaction—Satisfaction with agencywide applications based on results from the NRC employee survey.
3. Identify needs and provide necessary resources to deliver IT/IM services to all offices to help them take advantage of IT/IM capabilities to improve the effectiveness and efficiency of their operations.	3. Shared Data—Increase in amount of authoritative data made available for sharing in each major Federally identified line of business (e.g., financial management and human resources management) and for each NRC-specific line of business.
4. Strengthen the IT/IM governance framework to improve IT investment selection, control, and evaluation, and better integrate IT governance with the NRC's other planning and budgeting processes.	4. E-Gov Milestones—Percent of milestones completed as agreed to with OMB in the NRC's Electronic Government Implementation Milestone Plan.
5. Seek common solutions, reduce duplication, and promote sharing of data, systems, and service components across the agency.	5. OMB Exhibit 300 Scores—Percent of major IT investments that are rated as "acceptable" based on OMB evaluation of the NRC's Exhibit 300 submittal.
6. Influence Federal Government initiatives that are applicable to the NRC and quickly adopt Governmentwide IT solutions when they provide sufficient return on investment.	6. Enterprise Architecture Maturity—Maturity level of the NRC's enterprise architecture as assessed by the NRC using criteria set by the Government Accountability Office.
7. Strengthen IT project management.	
8. Increase staff awareness, proficiency, and innovation in applying IT/IM tools to strengthen individual and organizational performance	

The following section provides example means for executing each of the strategies that support the IT Applications goal, followed by a brief discussion of the measures for this goal.

Strategy 1—Before implementing new IT applications, evaluate and improve associated business processes, considering the needs of all process participants and using the most effective redesign approaches and technologies

In redesigning and improving IT applications, the NRC will focus on improving the process before automating, integrating across functions rather than accepting fragmented or duplicative solutions, and adapting to Governmentwide and commercially available approaches rather than opting for customization.

The NRC's approach to selecting the processes that need improvement and setting performance targets will include the following:

- use of the Performance Assessment Review Tool results and other objective measures of business process performance, such as unit cost per output, productivity, cycle time, and decrease in the number of touch points between processes
- mid- and senior-level NRC management involvement to identify target business processes and a schedule for improving them
- input from internal and external stakeholders involved in the processes
- selection criteria such as satisfaction with the process and support for strategic goals

The approach to implementing revised processes will include effective participation by internal and external stakeholders involved in the processes and a method for monitoring the progress of the process improvements.

Strategy 2—Apply IT/IM to meet high-priority business needs

The NRC will ensure that sufficient IT resources are applied to high-priority business needs such as those associated with new reactors, the high-level waste repository proceeding, and homeland security. Example means include strengthening the role of senior program executives in IT governance and budget processes and increasing IT project management support for high-priority areas.

Strategy 3—Identify needs and provide necessary resources to deliver IT/IM services to all offices to help them take advantage of IT/IM capabilities to improve the effectiveness and efficiency of their operations

For example, the NRC will increase assistance to smaller offices to aid them in obtaining IT/IM services, developing small applications in support of their needs, and using applications developed by other offices.

Strategy 4—Strengthen the IT/IM governance framework to improve IT investment selection, control, and evaluation, and better integrate IT governance with the NRC's other planning and budgeting processes

The NRC will continue to build on progress made to date in integrating the IT governance framework with the agency's planning, budgeting, and performance management process; its financial and human resources management processes and its program decisionmaking processes. The agency will also clarify roles and responsibilities and integrate, document, and fully implement IT/IM governance processes to effectively and efficiently deliver IT/IM business solutions, balancing compliance with service and efficiency.

Examples of processes covered by Strategy 4 include the following:

- project management methodology
- enterprise architecture
- capital planning and investment control for the NRC's IT portfolio

- IT/IM planning, budgeting, and performance management including OMB reporting

Strategy 5—Seek common solutions, reduce duplication, and promote sharing of data, systems, and service components across the agency

Example means for implementing this strategy include the following:

- identifying data owners, establishing central, authoritative data sources for use by all agency systems, documenting this in the NRC data reference model, and measuring the amount of shared data in major lines of business
- working with NRC business organizations to document and develop a future systems architecture through an agency transition plan that fully meets the business needs of NRC offices
- assisting NRC offices in leveraging existing and planned systems and data repositories by publishing a technical standards profile identifying all technology and data standards used at the NRC and their approved purpose

Also related to the goal of reducing duplication is Strategy 2 under Goal 4, IT Infrastructure (Section 4.4).

Strategy 6—Influence Federal Government initiatives that are applicable to the NRC and expeditiously adopt Governmentwide IT solutions when they provide sufficient return on investment

Examples of applicable initiatives are the human resources and financial management lines of business, electronic rulemaking, e-records management, electronic forms, case management, and geospatial initiatives.

Strategy 7—Strengthen IT project management

Example means include the creation of an IT project management function to provide training, assistance, and tools for use by IT project managers throughout the agency. Services may include establishing an IT project management community of practice, strengthening IT project management training, and providing assistance in such areas as capital planning and investment control, risk management, requirements analysis, business process improvement, cost estimation, work breakdown structures, and earned value management.

Strategy 8—Increase staff awareness, proficiency, and innovation in applying IT/IM tools to strengthen individual and organizational performance

Examples means include the following:

- increasing awareness of best practices in applying IT/IM tools to NRC business needs
- encouraging new ideas for the application of IT/IM to meet business needs
- identifying best practices in using agency-standard desktop tools such as email and calendars and establishing consistent agencywide approaches
- strengthening IT/IM skills
- involving training staff early to ensure that new software implementation plans address training needs
- reviewing training periodically to ensure that training is effective and responds to customer needs

Measures for Goal 2

The NRC is adopting six measures to monitor its progress in achieving the major elements of the IT Applications goal. The three measures related to data sharing, enterprise architecture, and E-Government place a strong emphasis on gaining efficiencies through standards, resource sharing, and collaboration. In addition, the agency will assess the performance of its IT applications through measures of satisfaction, availability, and OMB Exhibit 300 scores..

4.3 Goal 3: IT Security

IT Security: Protect NRC's information and information systems to ensure their integrity, confidentiality, and availability.

Strategies	Performance Measures
1. Strengthen the security controls that protect NRC's IT systems and information using an efficient and effective certification and accreditation process.	1. System Certification and Accreditation—Percent of major applications and general support systems that have been certified and accredited.
2. Update the NRC's IT security policy, standards, and training to address current Federal guidance and the changes in the threat environment.	2. Certification and Accreditation Process Effectiveness Rating of the NRC's certification and accreditation process based on the annual IG assessment.
3. Strengthen cyber security situational awareness and incident response.	3. Contingency Plan Testing—Percent of major applications and general support systems that have completed the annual test of their contingency plans.
4. Revise and clarify the NRC's program for handling sensitive unclassified non-safeguards (SUNSI) information, including incorporation of Governmentwide requirements for Controlled Unclassified Information (CUI).	4. Security Plan of Actions and Milestones (POAM) Process Effectiveness—Rating of the NRC's IT Security POAM process based on the annual IG assessment.
5. Develop an agencywide, long-term information security strategy.	5. Privacy Impact Assessments—Of the NRC systems containing personally identifiable information, the percent that have completed privacy impact assessments.
	6. IT Security Awareness Training—Percent of personnel with logon access to the NRC's IT infrastructure that have completed the annual IT and information security awareness training.
	7. Role-Based IT Security Training—Percent of identified staff who have completed the applicable role-based security training.
	8. Information Security Infractions—Reduction in information security infractions.

The following section provides example means for executing each of the strategies that support the IT Security goal, followed by a brief discussion of the measures for this goal.

Strategy 1—Strengthen the security controls that protect NRC's IT systems and information using an efficient and effective certification and accreditation process

The NRC will increase staffing and resources for Federal Information Security Management Act (FISMA) compliance activities and improve the effectiveness and efficiency of the IT certification and accreditation process. The revised process will promote a better understanding of enterprise-wide mission risks resulting from the operation of information systems; create more complete, reliable, and trustworthy information for authorizing officials—facilitating more informed security accreditation decisions; and improve the security of the NRC's information systems.

Strategy 2—Update the NRC's IT security policy, standards, and training to address current Federal guidance and the changes in the threat environment

The NRC will update and consolidate its existing IT security directives, rules, and practices to make them more useful to the staff and to incorporate recent changes in Federal guidance. The agency will continue to improve its annual IT and information security awareness training for personnel with logon access to the NRC's IT infrastructure, and introduce role-based IT security training for employees with specific IT roles, such as system administrators and information system security officers.

Strategy 3—Strengthen cyber security situational awareness and incident response.

The NRC will expand its staffing and improve the tools it uses for cyber security situational awareness and incident response.

Strategy 4—Revise and clarify the NRC's program for handling SUNSI, including incorporation of Governmentwide requirements for CUI

Example means include the following:

- Review and revise the current SUNSI policy.
- Develop a communications plan.
- Develop and deliver training.

Strategy 5—Develop an agencywide, long-term information security strategy

The NRC will develop, document, and implement an agencywide information security strategy that is responsive to Federal requirements and any unique requirements of the NRC's mission, systems, and information.

Measures for Goal 3

The NRC is adopting eight measures to monitor its progress in achieving the major elements of the IT Security goal. Three focus on IT systems security, one ensures progress in remediating any weaknesses based on the annual IG assessment, one relates to privacy data, two involve IT security training, and one measures the reduction in information security infractions..

4.4 Goal 4: IT Infrastructure

IT Infrastructure: Provide an IT/IM infrastructure that is secure, robust, reliable, and responsive to changing business needs.

Strategies	Performance Measures
1. Improve IT infrastructure planning. 2. Build shared services into the IT infrastructure to reduce costs of applications that require these services. 3. Improve IT infrastructure operational security. 4. Improve IT infrastructure delivery processes and services. 5. Increase IT infrastructure capacity, availability, and reliability to cost effectively meet business needs. 6. Provide IT infrastructure products and services that respond to the needs of a mobile workforce while continuing to ensure information security. 7. Strengthen IT infrastructure capabilities to accommodate agency business needs during emergencies. 8. Provide IT infrastructure capabilities to accommodate classified and safeguards information.	1. Availability—Percent of the time that key IT infrastructure services are available. 2. Staff Satisfaction—Average satisfaction score for IT infrastructure services based on applicable questions in the NRC employee survey. 3. Security Vulnerability Remediation Timeliness—Percent of identified security vulnerabilities that are addressed in accordance with the time limits specified in the NRC's security procedures.

The following section provides example means for executing each of the strategies that support the IT Infrastructure goal followed by a brief discussion of the measures for this goal.

Strategy 1—Improve IT infrastructure planning

This strategy includes improving the requirements analysis process for IT infrastructure upgrades in partnership with internal stakeholders to take into account such questions as the following:

- What emerging business needs do we need to support?

- What technologies are currently supported?

- What supported technologies need to be upgraded?

- What new technologies should we adopt?

- What is the impact on our IT architecture?

- What are the current service-level agreements?

- What service-level agreements need to be changed?

The NRC will use the outcome of the planning process to update a single, integrated technology roadmap as part of its enterprise architecture transition plan.

Strategy 2—Build shared services into the IT infrastructure to reduce costs of applications that require these services

Examples of shared services covered by this strategy include identity management (single sign-on), workflow, electronic meetings and collaboration, and secure communications. Another example is agency participation in the implementation of HSPD-12. Under HSPD-12, there will be a mandatory, Governmentwide standard for the identification issued by the Federal Government to its employees and contractors. This universal Government "smartcard" will be used to control physical and electronic access to Federal buildings, applications, and other resources to enhance security, increase Government efficiency, reduce identity fraud, and protect personal privacy.

Strategy 3—Improve IT infrastructure operational security

Example means for implementing this strategy include the following:

- providing automated security patch management and vulnerability assessment capability for agency servers and workstations
- expanding network intrusion detection capability

Strategy 4—Improve IT infrastructure delivery processes and services

Examples of potential improvements include the following:

- simplifying the process of deploying desktop technology for employees new to the NRC or new to a position
- improving service delivery, for example by expanding primary support hours for the Customer Support Center to include weekend support
- providing more flexibility in solutions to program needs

Strategy 5—Increase IT infrastructure capacity, availability, and reliability to cost effectively meet business needs

Example means include the following:

- improving Web-based multimedia capabilities to enable such uses as distance learning, broad access to Webinars, and access to nontextual records such as video, photo, and audio archives
- participating in the Federal initiative to implement IPV6, which expands Internet address spaces, providing the necessary basis for realizing a global information society and laying the ground for convergence between fixed and mobile computing as well as between data, voice, and video transmission through the Internet
- maintaining the NRC's video teleconferencing system with up-to-date equipment that meets the growing needs of the agency
- increasing the level of redundancy in the NRC's connection to the Internet to reduce risks from common-mode failures and single points of failure
- increasing IT infrastructure capacity in response to agency growth, the expanding use of nontextual media, and new business needs

Strategy 6—Provide IT infrastructure products and services that respond to the needs of a mobile workforce while continuing to ensure information security

Examples of means to implement this strategy are expanding remote access capability for NRC staff, expanding the use of wireless computing, and increasing support for mobile computing devices such as laptops, dockable workstations, and personal digital assistants. Because these new capabilities introduce additional risks, the NRC will implement new policies and controls to protect information privacy and security and provide staff with guidance on appropriate management and use.

Strategy 7—Strengthen IT infrastructure capabilities to accommodate agency business needs during both nuclear and nonnuclear emergencies

Examples include providing necessary IT/IM capabilities in the NRC Operations Center; increasing the involvement of IT/IM staff in continuity of operations planning; identifying vital records; documenting procedures, roles, and responsibilities for IT/IM staff during incidents and emergencies; and increasing IT/IM staff participation in exercises, drills, tests, and other practice activities.

Strategy 8—Provide IT infrastructure capabilities to accommodate classified and safeguards information

Examples are implementing a secure IT infrastructure, secure videoteleconferencing, and participating in related Federal initiatives such as the Homeland Secure Data Network and secure Federal digital signatures.

Measures for Goal 4

The NRC has chosen three measures to assess agency performance in meeting the IT Infrastructure goal. The results of these measurements will guide work under the IT infrastructure strategies and focus activities on areas needing improvement.

4.5 Goal 5: IT/IM Human Capital

IT/IM Human Capital: Recruit, develop, and retain a highly capable IT/IM workforce with the competencies needed to support NRC's goals and objectives.	
Strategies	**Performance Measures**
1. Improve IT/IM workforce identification and more effectively utilize existing IT/IM skills assessment and reporting tools to support agency staffing requirements, identify skill gaps, and respond to overall IT/IM Federal workforce trends. 2. Offer robust NRC IT/IM professional development programs to address significant skill gaps. 3. Fully utilize available Federal competitive compensation and workforce flexibilities as well as specialized recruitment and development programs to attract and retain top level IT/IM talent.	The measures for the IT/IM Human Capital goal will be included as part of overall agency human capital measures in the next version of the NRC Human Capital Plan.

The following are example means for executing each of the strategies that support the IT/IM Human Capital goal.

Strategy 1—Improve IT/IM workforce identification and more effectively utilize existing IT/IM skills assessment and reporting tools to support agency staffing requirements, identify skill gaps, and respond to overall IT Federal workforce trends

IT/IM workforce planning is an ongoing process. Critical steps in workforce planning include accurately identifying and classifying the IT workforce, summarizing the IT skill gaps of these individuals, and understanding workforce trends, such as pending retirements, so that strategies can be implemented to meet future challenges.

Strategy 2—Offer robust NRC IT/IM professional development programs to address significant skill gaps

The activities supporting this strategy will focus on professional development and training programs to close critical skill gaps identified through the NRC's strategic workforce planning systems and other means, with an initial emphasis on IT security and IT project management skills.

Strategy 3—Fully utilize available Federal competitive compensation and workforce flexibilities as well as specialized recruitment and development programs to attract and retain top level IT/IM talent

The NRC must fully utilize a variety of recruitment and retention strategies and tools to attract a diverse, highly qualified applicant pool for IT/IM positions. Examples could include offering retention incentives to individuals with critical skills and using recruitment and development programs such as the Nuclear Safety Professional Development Program and the CoOp program. In addition, the increasing Federal competition for IT/IM talent, in combination with looming retirements, makes it essential for the NRC to stay attuned and responsive to exit rates of IT/IM employees.

4.6 Goal 6: IT/IM Customer Service

IT/IM Customer Service: Achieve and sustain a high level of satisfaction with agencywide information services.	
Strategies	**Performance Measures**
1. Institutionalize a friendly IT/IM customer service culture. 2. Improve the effectiveness and efficiency of help services for the staff and the public. 3. Increase awareness of existing NRC IT/IM customer services and processes.	Measures for this goal are being addressed using staff and stakeholder satisfaction measures under Goals 1, 2, and 4. See Goal 1, measures 2 and 3; Goal 2, measure 2; and Goal 4, measure 2.

The following are example means for executing each of the strategies that support the IT/IM Customer Service goal.

Strategy 1—Institutionalize a friendly IT/IM customer service culture

As part of this strategy, the NRC intends to proactively involve customers in improvements to IT/IM processes, services, and communications. Means for identifying improvement needs may include surveys, best practice reviews, and stakeholder suggestions.

Strategy 2—Improve the effectiveness and efficiency of help services for the staff and the public

Example means for implementing Strategy 2 include the following:

- establishing integrated, tiered support for all external users
- expanding and improving IT/IM help services for internal users, including integration of all internal assistance resources
- strengthening help desk services related to the use of supported software
- increasing awareness of IT/IM customer services and processes

Strategy 3—Increase awareness of existing NRC IT/IM customer services and processes

The NRC will develop and execute a communications plan to help internal and external stakeholders become more aware of existing IT/IM customer service resources.

5. IT/IM Strategic Planning and Performance Measurement Process

5.1 Description of the IT/IM Planning and Performance Measurement Process

In FY 2005, the NRC began a new IT/IM strategic planning process designed to more effectively involve internal stakeholders and to better integrate with other agency planning and budgeting processes. Figure 2 depicts this process. The process has three steps, each of which involves three stages. During the first step, the vision and goals are developed; during the second step, the strategies and measures are developed; and during the third step, the NRC IT/IM Strategic Plan is produced. At each step, OIS, the NRC's centralized IT/IM support office, develops an initial planning product. Next, the IT/IM Strategic Planning Group, composed of senior managers representing the major NRC programmatic, regional, and support offices, reviews this product. At the final stage of each step, the product receives the concurrence of the IT Senior Advisory Council, composed of the directors of the NRC's major programmatic and support offices as well as a regional administrator who represents the NRC's four regional offices. At the end of the planning process, the NRC publishes the vision, goals, strategies, and measures in the NRC IT/IM Strategic Plan, which must be approved by the CIO and EDO.

Figure 2
IT/IM Strategic Planning Process

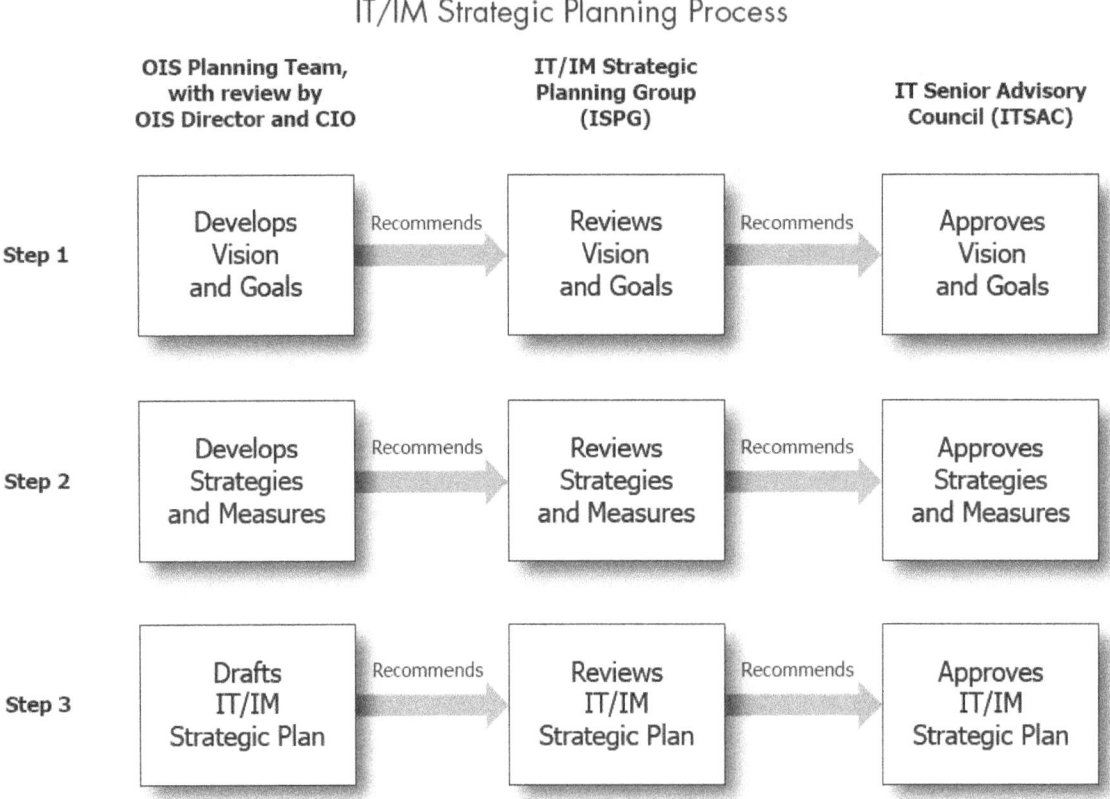

Upon approval by ITSAC, the IT/IM Strategic Plan goes through the final concurrence process, and is approved by the CIO and EDO for posting at the NRC's public Web site.

Selected IT/IM measures from this plan are tracked at the agency level in the NRC's Performance Budget. The other measures are tracked in annual office operating plans and in the performance plans of appropriate senior executives. Under the leadership of the CIO and Deputy CIO, the agency monitors, tracks, and trends the progress of these measures over time with the goal of continuous improvement, conducting regular performance reviews to assess progress. The agency will use the results of these performance reviews to adjust the strategies, means, and measures where appropriate.

5.2 Relationship to Other Planning and Performance Measurement Processes

The NRC framework for performance-based management is the Planning, Budgeting, and Performance Management process that was established in January 1998 and updated in July 2002. This process implements the Government Performance and Results Act, which requires the transmittal of a Strategic Plan, Performance Budget, and Performance Report to Congress. The NRC has designed the new IT/IM strategic planning process to be an integral part of the agency's Planning, Budgeting, and Performance Management process.

Annually, the Commission provides guidance on the agency's outcome-based performance measures, which indicate the level of success needed to achieve the agency's goals. The performance measures form the basis for the NRC to develop key planning assumptions, which identify major program drivers that would significantly influence the NRC's work activities and resource requirements. For each major activity, the agency identifies the major program outputs and output-based measures needed to achieve the outcome-based performance measures, taking into consideration the key planning assumptions. The NRC also identifies and prioritizes planned activities, including those for IT/IM, needed to achieve the outputs in each major activity and then prioritizes them based on their contribution to the goals. Lastly, the NRC determines the resource requirements needed to achieve each planned activity, which form the basis for developing the agency's budgetary requests for each program area. Each of the NRC's Performance Budget review levels takes into consideration those factors described above in relating outcome-based and output-based performance measures to resources in making budget recommendations and decisions.

Figure 3 illustrates the manner in which the agency's Strategic Plan and the IT/IM Strategic Plan drive the annual performance management and budgeting processes. The CIO leads the annual performance review for the agency's IT/IM program with the participation of the IT Senior Advisory Council. The results of the performance review are used to help set IT/IM performance targets for upcoming budget cycles.

The update cycles for the agency Strategic Plan and IT/IM Strategic Plan are now synchronized so that major updates occur in tandem, helping to maintain alignment between the two plans.

Figure 3
The NRC's Annual IT/IM Performance Management Review and Its Relationship to the Budget Process (Example for FY 2009)

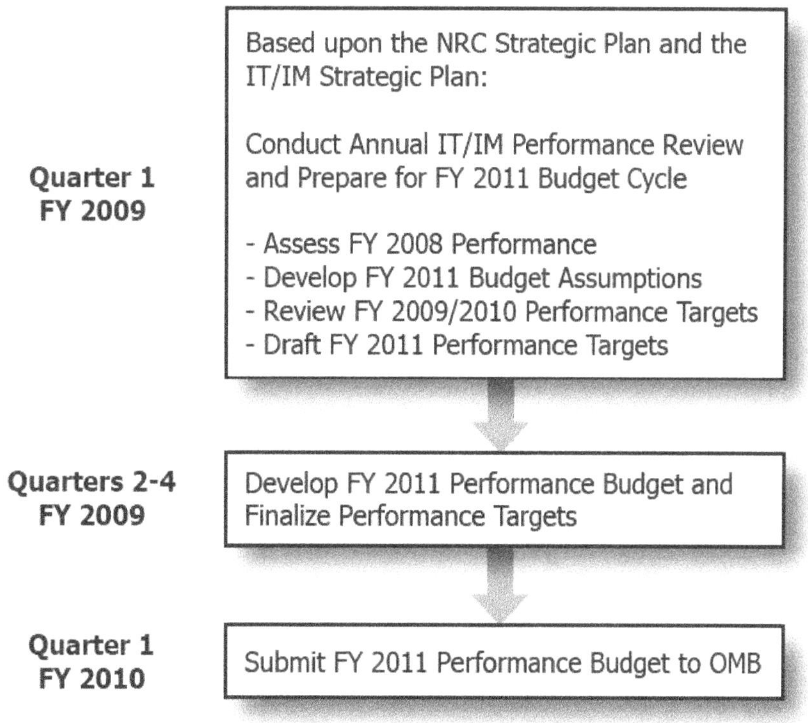

Quarter 1 FY 2009	Based upon the NRC Strategic Plan and the IT/IM Strategic Plan: Conduct Annual IT/IM Performance Review and Prepare for FY 2011 Budget Cycle - Assess FY 2008 Performance - Develop FY 2011 Budget Assumptions - Review FY 2009/2010 Performance Targets - Draft FY 2011 Performance Targets
Quarters 2-4 FY 2009	Develop FY 2011 Performance Budget and Finalize Performance Targets
Quarter 1 FY 2010	Submit FY 2011 Performance Budget to OMB

5.3 Roles and Responsibilities for IT/IM Strategies and Measures

The NRC's CIO has overall responsibility for the IT/IM strategic planning process. In addition, the NRC has assigned a senior executive sponsor for each strategy and measure in this Strategic Plan. That senior executive is responsible for working with representatives of other interested NRC organizations to develop tactical plans for implementing the strategy or measure. The responsible senior executive is also responsible for tracking the strategy or measure across the agency and reporting results during the annual IT/IM performance review. Directors of internal stakeholder organizations are responsible for tracking appropriate supporting initiatives and measurement inputs in their annual operating plans. The Deputy CIO directs the overall implementation of the plan, ensures coordination between strategy sponsors, and tracks progress of the measures for use in performance reviews.

5.4 Relationship to Federal Electronic Government Strategy

In December 2002, President Bush signed the Electronic Government Act (E-Gov). This act defined Governmentwide strategies for achieving better service to citizens and reducing IT costs for the Government (the annual Federal IT budget is $65 billion). The NRC's IT/IM strategies, as reflected in this plan, fully support the Government's E-Gov program.

Governmentwide strategies for E-Gov have focused on (1) the development and implementation of Presidential Priority Initiatives (PPIs), (2) the development and implementation of the Line of Business (LoB) programs, (3) the establishment of a Federal Enterprise Architecture and of enterprise architectures at all Federal agencies, and (4) information security. A key characteristic of the E-Gov program is its focus on common needs and services shared by agencies across the Government (e.g., payroll, human resource management, support for citizen access to Government services, records management, training, etc.). E-Gov programs largely focus on gaining efficiencies and improving the effectiveness of administrative and support functions. This Strategic Plan includes the NRC's strategies for further alignment with E-Gov under Goal 2, IT Applications (Section 4.2).

The E-Gov program has identified 25 PPIs and 9 LoB[1] programs as of April 2006. On an ongoing basis, the NRC reviews E-Gov initiatives to assess their relevance to the agency and, as part of the CPIC process, screens all IT investments to ensure that they do not overlap with any E-Gov program. As a result of these reviews, the NRC is participating in 16 of the 25 PPIs and 6 of the 9 LoB programs. The NRC has established the position of Senior Advisor for Electronic Government and has institutionalized procedures to ensure that IT/IM investments comply with E-Gov guidance and objectives. These actions provide for management oversight and, in addition, the NRC maintains an E-Gov action plan with progress periodically reported to NRC management, OMB, the NRC's IG, and GAO (on request).

E-Gov also requires agencies to focus on their programs for information security and enterprise architecture and to adopt new technologies and more efficient ways of doing business. Goal 3, IT Security (Section 4.3) comprises the NRC's primary strategies for information security. Related strategies also appear under Goal 1, Information (Section 4.1) and Goal 4, IT Infrastructure (Section 4.4). A strategy and measure under Goal 2, IT Applications (Section 4.2), addresses enterprise architecture. The NRC integrates new technologies that are being implemented across the Federal Government (e.g., IPV6 and the Federal identification card required by HSPD-12) consistent with Governmentwide guidance and available resources. The NRC's budget submittals provide information on the implementation of such technologies at the project level.

---

1. While LoBs and PPIs are both cross-cutting initiatives, the LoB program is much broader in scope and contains various PPIs. For example, an LoB addresses an entire area, such as human resources, while the e-training PPI is one of 10 human resource functions contained in the human resources LoB.

APPENDIX

Relationship of IT/IM Strategic Plan to the Agency Strategic Plan

This appendix shows the relationship between specific strategies, means, and activities for each goal and organizational excellence objective in the NRC Strategic Plan for Fiscal Years 2008—2013 and the six IT/IM strategic goals. The numeric entries under each IT/IM goal indicate an associated strategy that supports the corresponding element of the NRC Strategic Plan.

Goals and Selected Strategies, Means, and Activities in the NRC Strategic Plan for Fiscal Years 2008—2013	IT/IM Goals with Relevant Strategies					
	Infor-mation	IT Applications	IT Security	IT Infra-structure	IT/IM Human Capital	IT/IM Customer Service
SAFETY GOAL						
Continue to oversee the safe operation of existing plants while preparing for and managing the review of applications for new power reactors. (Strategy 2)	1.3	2.2				
Effectively respond to events at NRC-licensed facilities and other events of national interest, including maintaining and enhancing the NRC's cr tical incident response and communication capabilities (Strategy 9)			3.3	4.7		
SECURITY GOAL						
Use relevant intelligence information and security assessments to maintain realistic and effective security requirements and mitigation measures (Strategy 1)	1.6		3.2	4.8		
Share security information with appropriate stakeholders and international partners (Strategy 2)	1.6			4.8		
Control the handling and storage of sensitive security information, and the communication of information to licensees and Federal, State, and local partners. (Strategy 4)	1.6		3.1, 3.4	4.3		
Enhance the programs for control of the security of radioactive sources and strategic special nuclear material commensurate with their risk, including enhancements required by the Energy Policy Act of 2005. (Strategy 7)		2.2				
Collaborate with the Department of Energy, Department of Homeland Security, and other agencies and state governments to develop and implement a national registry of radioactive sources. Improve the controls on high-risk radioactive materials, including enhancements required by the Energy Policy Act of 2005 and recommended by the Task Force on Radiation Source Protection and Security, to prevent their harmful use (Means supporting Strategies 1, 4, 6, and 7)		2.2				
Identify and develop key information technology investments, including secure electronic document and records management capabilities, that will enhance the storage, handling, and communication of sensitive security information both within and external to the agency (Means supporting Strategy 4)	1.6	2.2		4.8		

Goals and Selected Strategies, Means, and Activities in the NRC Strategic Plan for Fiscal Years 2008–2013	IT/IM Goals with Relevant Strategies					
	Infor-mation	IT Applications	IT Security	IT Infra-structure	IT/IM Human Capital	IT/IM Customer Service
OPENNESS (Organizational Excellence Objective)						
Enhance the awareness of the NRC's independent role in protecting public health and safety, the environment, and the common defense and security (Strategy 1)	1.5					
Provide accurate and timely information to the public about NRC's mission, regulatory activities, and performance and about the uses of, and risks associated with, radioactive materials (Strategy 2)	1.3, 1.4					
Provide for fair, timely, and meaningful stakeholder involvement in NRC decisionmaking without disclosing classified, safeguards, proprietary, and sensitive unclassified information (Strategy 3)	1.1, 1.3		3.4	4.3		
Communicate about NRC's role, processes, activities and decisions in plain language that is clear and understandable to the public (Strategy 4)	1.4					
EFFECTIVENESS (Organizational Excellence Objective)						
Use state-of-the-art technologies and risk insights to improve the effectiveness and realism of NRC actions with a goal of continuous improvement (Strategy 1)		2.8				
Continue to improve the NRC's regulatory and communication programs (Strategy 7)	1.3	2.2				
Achieve efficiencies in the licensing process that enable the safe and secure use of nuclear material (Strategy 8)	1.2	2.1, 2.2, 2.3, 2.5, 2.8		4.2		
OPERATIONAL EXCELLENCE (Organizational Excellence Objective)						
Improve support services to make them more efficient and make it easier to accomplish agency goals (Strategy 3)	1.2	2.4, 2.5		4.1, 4.2, 4.4, 4.5, 4.6		6.1, 6.2
Manage information and employ information technology to improve the productivity, effectiveness, and efficiency of agency programs and enhance the availability and usefulness of information to all users inside and outside the agency (Strategy 4)	All	All	All	All		All
Use innovative strategies to recruit, develop, and retain a high-quality, diverse workforce (Strategy 5)					5.3	
Sustain a learning environment that provides continuing improvement in performance through knowledge management, performance feedback, training, coaching, and mentoring (Strategy 7)	1.2	2.7, 2.8			5.1, 5.2	
Provide accurate, timely, and useful financial information to agency managers for effective decisionmaking (Strategy 9)	1.1	2.2